STARK LIBRARY OCT - - 2022

DISCARD

SEA TURTLES

by Emma Bassier

Cody Koala
An Imprint of Pop!
popbooksonline.com

abdobooks.com

Published by Pop!, a division of ABDO, PO Box 398166, Minneapolis, Minnesota 55439. Copyright © 2020 by POP, LLC. International copyrights reserved in all countries. No part of this book may be reproduced in any form without written permission from the publisher. Pop!™ is a trademark and logo of POP, LLC.

Printed in the United States of America, North Mankato, Minnesota

052019
092019

THIS BOOK CONTAINS RECYCLED MATERIALS

Cover Photo: iStockphoto
Interior Photos: iStockphoto 1, 5 (top), 5 (bottom left) 5, 5 (bottom right), 7, 8, 11, 13, 14, 15, 17 (top), 17, (bottom left), 17 (bottom right), 18, 19, 21
Editor: Meg Gaertner
Series Designer: Sophie Geister-Jones

Library of Congress Control Number: 2018964500
Publisher's Cataloging-in-Publication Data
Names: Bassier, Emma, author.
Title: Sea turtles / by Emma Bassier.
Description: Minneapolis, Minnesota : Pop!, 2020 | Series: Ocean animals | Includes online resources and index.
Identifiers: ISBN 9781532163418 (lib. bdg.) | ISBN 9781644940143 (pbk.) | ISBN 9781532164859 (ebook)
Subjects: LCSH: Sea Turtles--Juvenile literature. | Marine turtles--Juvenile literature. | Sea turtles--Behavior--Juvenile literature. | Water reptiles--Juvenile literature.
Classification: DDC 597.72--dc23

Hello! My name is
Cody Koala

Pop open this book and you'll find QR codes like this one, loaded with information, so you can learn even more!

Scan this code* and others like it while you read, or visit the website below to make this book pop.

popbooksonline.com/sea-turtles

*Scanning QR codes requires a web-enabled smart device with a QR code reader app and a camera.

Table of Contents

Chapter 1
Ocean Reptile 4

Chapter 2
Inside the Shell 6

Chapter 3
Daily Life 12

Chapter 4
Life Cycle 16

Making Connections 22
Glossary. 23
Index 24
Online Resources 24

Chapter 1

Ocean Reptile

Sea turtles are ocean **reptiles**. There are seven different kinds of sea turtles. They live in ocean waters around the world.

Watch a video here!

Chapter 2

Inside the Shell

A sea turtle has a large, hard shell. The shell protects the sea turtle from other animals. The shell can be green, brown, red, or yellow.

Learn more here!

Sea turtles have **flippers**. The front flippers are like paddles. They help the turtle move in the water. The back flippers are shorter. They help the turtle **steer**.

> Unlike land turtles, sea turtles cannot pull their flippers or heads inside their shells.

Sea turtles breathe air. But they can stay underwater for a long time. Green sea turtles can be underwater for up to five hours.

Chapter 3

Daily Life

Sea turtles live alone. They dive deep to get food. When sea turtles are not feeding, they come to the water's surface. They **bask** in the sun.

Learn more here!

13

Some sea turtles eat plants. Other sea turtles eat animals such as jellyfish.

Sea turtles have a great sense of smell. It helps them find food in the water.

Chapter 4

Life Cycle

Female sea turtles find a beach. They dig a hole. They lay 50 to 200 eggs. Then they cover the eggs with sand. After several months, the eggs hatch.

Complete an activity here!

Baby turtles dig to the surface. They head to the ocean. They swim for days.

They need to get to deep water. The deep water helps them stay safe.

Many animals eat baby sea turtles. More than 90 percent of the babies will be eaten. But sea turtles that survive childhood can live for decades.

> Most females return to the same beach where they hatched.

Making Connections

Text-to-Self

Sea turtles breathe air. But some can stay underwater for hours. How long can you hold your breath?

Text-to-Text

Have you read other books about reptiles? What new things did you learn in this book?

Text-to-World

Sea turtles have shells that protect them. What other animals have shells?

Glossary

bask – to lie in the light of the sun.

female – a person or animal of the sex that can have babies or lay eggs.

flipper – one of four limbs on the sides of a sea turtle that help it swim.

reptile – a type of animal that lays eggs and has scales, rough skin, or a shell.

steer – to control the direction of movement.

Index

basking, 12
breathing, 10
eggs, 16
females, 16, 20

flippers, 8, 9
food, 12, 14–15
reptiles, 4
shells, 6, 8, 9

Online Resources

popbooksonline.com

Thanks for reading this Cody Koala book!

Scan this code* and others like it in this book, or visit the website below to make this book pop!

popbooksonline.com/sea-turtles

*Scanning QR codes requires a web-enabled smart device with a QR code reader app and a camera.

3 1333 05174 1450